Keep Bees
without Fuss *or* Chemicals

a detailed guide to the efficient
nent

by

Joe Bleasdale

Keep Bees without Fuss or Chemicals
© 2012 Joe Bleasdale

Little Orchard, Hardway, Bruton, BA10 0LN
joebleasdale@btinternet.com
e-Book at
amazon.co.uk/dp/B0073RV1YM

ISBN 978-1-908904-14-0

Published by Northern Bee Books, 2012
Scout Bottom Farm
Mytholmroyd
Hebden Bridge
HX7 5JS (UK)

Design and artwork
D&P Design and Print
Worcestershire

Printed by Lightning Source, UK

Keep Bees
without Fuss *or* Chemicals

a detailed guide to the efficient
hive management

by

Joe Bleasdale

Addendum to 'Checking for Brood Diseases', page 28:

In the UK European Foulbrood is also notifiable.

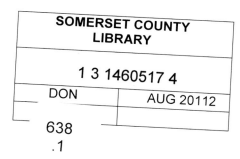

Northern Bee Books

Contents

Keep Bees without Fuss or Chemicals

A detailed guide to efficient hive management for new beekeepers, also for those established beekeepers whose hives suffer inexplicable colony losses, such as winter deaths, failing queens, paralysis etc. In my view weak colonies can be attributed to the use of chemicals and medicines in the hive, combined with unnecessary manipulation and disturbance of the brood.

This book gives the guiding principles and the step-by-step descriptions of the essential operations for efficient beekeeping. There are many practices that are described in other books and journals on beekeeping that are not covered, because I consider them to be fussy and unnecessary: certainly not to be undertaken by beginners. Such practices in my view cause disturbance in the hive and significantly weaken the colony. A hive of bees should be thought of as a single organism and its integrity not assailed by fiddling operations.

Research on the deleterious effect of the various chemicals used to control the varroa mite on the viability of queens and drone semen has reinforced my decision to abandon the use of such for the past 12 years.[1]

Acknowledgements

I would like to thank the following, who have given permission for the use of their photographs:

1. Gary Hutchinson of Informed Farmers, New South Wales Australia. Web site: http://informedfarmers.com/livestock/honey-bees/

2. Paul Brennan of Limerick Beekeepers in Eire, an organisation that is committed to advance the use of the native Irish Black Bee.
   Web site: http:/limerickbees.net/

3. Dr. Mary Coffey, Teagasc, Oakpark Research Centre, Carlow

---

[1] Lisa Marie Burley:

The Effects of Miticides on the Reproductive Physiology of Honey Bee

(*Apis mellifera L.*) Queens and Drones

http://scholar.lib.vt.edu/theses/available/etd-08162007-092313/unrestricted/lmburley.pdf

Introduction

This book is for those who would like to keep bees, but think that they may lack the time and resources.

Beekeeping is indeed an absorbing activity and time spent observing these marvellous and productive insects can be instructive and fulfilling. Many people have become interested in beekeeping recently, having heard and read about the plight of bees in the news and on television programmes. Beekeeping has become a fashionable hobby, but a large proportion of enthusiastic beginners have been discouraged, following losses of their colonies from various causes: disease, swarming, starvation, attacks by wasps and robbing by bees from other hives, and that modern affliction, the varroa mite. Other reasons for giving up have been poor seasons with no honey, lack of confidence in their ability to manipulate a hive, and so on.

I have tried to keep the book simple. The essential operations that the beekeeper should do over the year are described in detail, step by step. I have deliberately omitted many traditional beekeeping practices that evolved in the past two centuries with the stated aim of preventing swarming. One such was to open the hive every ten days or so in summer, to examine the brood frames and look for queen cells and destroy them. In my view this is futile: akin to cutting off the heads of dandelions in the hope that they will not set seed. Bees intent on swarming will swarm anyway, and all the beekeeper achieves is unnecessary disturbance of the bees, an angry hive and a reduction of the honey crop. After all, swarming is the natural way that bees reproduce and evolve. Why suppress this instinct?

I do not follow the current practice of treating the hive with chemicals, to combat various diseases and the dreaded varroa mite. I have found they are ineffective and weaken the bees. Over the past 20 years, British beekeepers have been feeding a variety of medicines to their bees, some of which have been effective up to a point, but varroa have evolved to become resistant. In a similar way that our hospitals have produced superbugs by the indiscriminate use of antibiotics, beekeepers have bred resistant varroa, when they should have been breeding resistant bees.

My approach has been to open up the hive only when necessary. Much can be learned about the state of a hive by close observation of the hive entrance and the bees flying in and out. Pollen on the legs of bees entering will indicate a laying queen. The debris on the ground in front of the hive entrance will indicate removal of old pollen and comb and dead bees. Bees crawling on the ground and unable to fly may indicate disease or the presence of varroa.

By quietly lifting the roof and looking over the top of the crown board without delving into the brood box and letting cold air chill the brood, one can check whether the hive is crowded and needs another super on top, or whether it is about to swarm.

In this book I will show you how to keep bees with a minimum of fuss and expense, to enjoy an absorbing and productive hobby. The practices I describe are based on over thirty years experience with my bees in the counties of Hampshire and Somerset. There is no prescriptive rule for beekeeping that will guarantee success, but there are some principles that will guide the beekeeper on what actions to take in given circumstances. These I have described in this book.

Why keep bees?

Beekeeping is an enjoyable and absorbing hobby and there is always something new to learn and observe in these fascinating and useful insects. You will benefit from the products of the hive: delicious and healthy honey, beeswax that can be used for all manner of things, and other products that are used in cosmetics and medicines. You will meet fellow beekeepers and make new friends who have a like-minded regard and love of nature. Your bees will benefit the local environment by pollinating fruit, seeds and berries to produce food for people and wild life. You will develop a better understanding of nature and help make the world a better place.

Who should keep bees?

To keep bees you should be reasonably fit and healthy. You should be even-tempered and calm when manipulating the hive, and not the type of person to be flustered or panic when a cloud of bees is flying around you. Nor should you be careless or foolhardy, for it is bad for you and the bees if you are badly stung because you failed to make your bee suit bee-proof, or opened the hive in unsuitable weather and without using the smoker. You should be observant, noting the behaviour of the bees at the hive entrance, what they are bringing in to the hive: nectar, pollen; or taking out of the hive: dead bees, debris of all sorts when they "spring clean", and drones which are expelled at the end of summer.

Stings

Though every precaution should be taken to prevent stings, a few stings during the season are almost inevitable. They will hurt to begin with, but most people will find that they become less painful as the season progresses. Despite all precautions, if the bees become very angry when the hive is open, then close it up quickly and calmly and leave the hive alone for the day. A few stings to the hands can be tolerated, but a sting to the head, particularly near the eyes or mouth, can be serious. Therefore it is essential to have bee-proof headgear, and wear it at all times when at the hive.

Stings may be treated with a squirt of proprietary spray such as Wasp-eze aerosol, and if very painful take an anti-histamine pill and put an ice cube on the affected area. If the sting is still attached to your skin, remove it by scraping it off with a fingernail, because it will continue to inject venom and attract other bees to the spot through the odour it emits. Rarely, some people may go into anaphylactic shock. Some beekeepers become sensitised after frequently being stung, and as a result may become prone to anaphylactic shock. That is a good reason for not fussing with the bees, keeping manipulations to the minimum and having bee tight and clean clothing. The symptoms and advice for treating anaphylactic shock can be found on a suitable web site such as the British Beekeepers Association, link:

http://www.bbka.org.uk/help/stings/anaphylactic_shock

Thus you are advised to join the local branch of your national beekeepers association, and work on your hive with an experienced beekeeper to help you, at least until you become confident and competent.

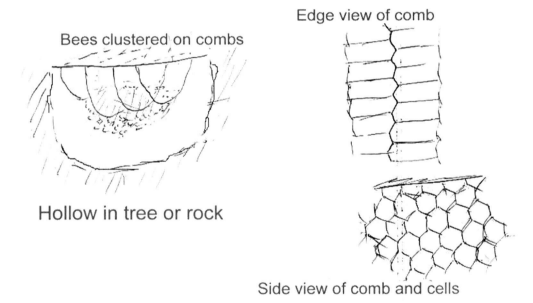

Bees clustered on combs

Edge view of comb

Hollow in tree or rock

Side view of comb and cells

Bees and comb in natural hive

The Natural Beehive

Honeybees are social insects and live in colonies of up to 80,000 bees. A colony will occupy hollow spaces such as hollow tree trunks, empty chimneys, and cavities in rocks or walls. The main requirement is that the colony is protected from the weather and can defend itself against predators large and small: insects, birds and mammals. Also there should be enough space to accommodate the bees and their stores – at least a cubic foot. Such suitable places will be sought by a swarm, which originates from an established hive. The swarm consists of between ten to forty thousand bees. They will have ingested honey to store in their stomachs, some of which they convert into wax through glands in their abdominal segments. They cling to the roof of their new home and within the warmth of the centre of the cluster will cooperate in moulding the wax into comb – see figure opposite.

They will start with a single comb and build parallel combs in a vertical plane running either side at about 1.4 inch separation. Each comb is made from a central sheet with hexagonal cells on each side. The cells are used to store the bees food: honey and pollen, and to raise the brood: eggs, larvae and pupa. When raising brood, the bees must maintain the temperature inside the cluster where the brood is developing at about 34 –35 degrees centigrade: nearly the same as our own body temperature. This is an important point to bear in mind when the beekeeper is managing the hive, and a good reason for only opening the hive when necessary and then in warm weather: - don't fuss with the bees!

Lifecycle of the Honeybee

There are three classes of honeybee in a hive: queen, drone and worker. The interval between the laying of the egg and the emergence of the bee from the pupal state is fixed for each class of bee. That is why the colony tries its utmost to maintain the temperature around the brood at a constant level, whatever the outside weather conditions, and why it is harmful to open the hive in cold or windy weather. Hence you should only open the hive to look in the brood chamber for a necessary purpose. These are the development times in the lifecycle from the laying of the egg to the emergence of the young insect from the pupa:

Queen:	16 days
Drone	24 days
Worker	21 days

Queen

In a well-regulated hive, the queen is the only one who lays eggs, one to a cell. The queen can live for up to six or seven years. Within the first 10 or so days after hatching, she will make up to a dozen or so mating flights within a few days, flying away from the hive to mate with drones from other hives. This will provide her with enough sperm for the rest of her life, to fertilise nearly a million eggs during the coming seasons. There is generally only one queen in a hive, for when a queen hatches she will seek out other virgin queens and they will fight to the death, using their sting to kill their rivals, either newly hatched queens or those still in their cells. During the spring and summer, a vigorous queen could lay up to a thousand eggs a day. She will reduce her rate of laying in autumn and stop laying in winter.

A queen is raised from a fertilised cell worker egg, where the cell is enlarged and the larva is fed "Royal Jelly" a milky food, throughout its larval stage, unlike the worker larvae, which are only fed Royal Jelly for the first 3 days, then honey and pollen.

Queen surrounded by her retinue of workers.

Drones

The drones are male bees who will fertilise a queen on her mating flight and then die. Typically there are around a hundred drones in a hive during the summer, so only a few manage to mate. At the end of summer, they are ejected from the hive by the worker bees, since there is no longer a need for them. Drone bees are raised from unfertilised eggs, consequently they have the same genes as their mother, unlike the worker bees, whose genes are from one of a dozen or so drones as well as the mother. This is a point to note when attempting to breed desirable traits: good drones flying in a mating area are as important as selecting good queens for propagation. Drone cells are larger than worker cells and are usually nearer the outside edges of the comb.

worker, drone and queen (marked)

Workers

The worker bees are raised from fertilised eggs. They do all the work necessary to maintain the hive and raise the brood of new bees. In an average hive in summer, there are about fifty thousand workers. In summer they live for eight to twelve weeks, but those raised in autumn will survive the winter until spring, when the queen starts laying again. In winter, the population of the hive will dwindle through natural mortality to about twenty thousand bees. They will form a close cluster on the combs in the centre of the hive, huddling together to keep cosy and warm. They gradually consume their stored honey, expending just enough energy to maintain a minimum survivable temperature inside the cluster throughout the cold weather. If they have consumed all their honey, they will starve and the whole colony will die. If the weather is mild and sunny, some bees will fly out for a short period to excrete, but will return to the cluster.

Though worker bees are female, in a well regulated hive they are inhibited from egg laying by a pheromone emitted by the queen: "queen substance". This is distributed throughout the hive by mutual grooming and feeding. If the hive is queenless for over two or three months, some worker bees may start laying, but since the eggs are not fertilised they will become drones. Such a hive is doomed to fail.

Detailed studies have found that the glands of workers develop and then atrophy, so that they perform the following duties at progressive stages in their life:

"nursing" – feeding the queen and young larvae with Royal Jelly; and the older larvae with diluted honey or nectar;

"housekeeping" duties – cleaning the cells and the hive;

storing pollen and nectar from the foragers in the combs;

converting nectar into honey by evaporation and adding enzymes;

wax making and comb building;

guarding the hive;

finally, foraging for nectar, pollen, water and propolis.

Despite this progression, they will vary their duties if the circumstances change in the colony, for example to defend against a threat to the hive, or to collect water in a drought. There is no unionised job demarcation in a beehive!

Senses and Communication

The management of bees should take account of the way bees sense their environment and communicate with each other.

Sight

Like all insects, bees see through compound eyes. They can see light of shorter wavelength than we can: into the ultraviolet spectrum, so flowers which may appear to us to have one colour could appear multi-coloured and patterned to bees. They can also sense the direction of the sun from the polarisation of light in the sky, which helps them to navigate when foraging.

Scent

Bees detect odour through their two antennae, which direct them to flowers with nectar and pollen. They can also detect scents generated by glands in their own bodies, chemical signals called pheromones, which are used to communicate the state within the colony, and alert the colony of threats and actions that should be taken for its well-being: actions such as defending the hive against intruders, raising new queens, deciding to swarm. The queen herself emits a pheromone called "queen substance" that is carried round the hive from bee to bee and prevents the worker bees from laying eggs. The hive can sense the presence of a good queen, an old or failing or missing queen, and will act accordingly by raising queen cells from young larvae, to replace the queen.

Sometimes the beekeeper can smell the pheromones given off by the bees, in particular the alarm pheromone, which is said to have the scent of banana oil. That will be a sign for the beekeeper to cease operations and close up the hive! Another distinctive smell is given off when bees wish to indicate the hive entrance to other bees that may be new foragers, unsure of its position. This smell is quite strong when a swarm is taken or when it is first entering the new hive. Bees align themselves with the entrance and lift their tails and fan the pheromone from scent glands in their tails. It is quite a pleasant smell, detectable to the human nose.

The beekeeper should be careful to remove a sting from the skin or cloth as soon as possible, for the sting gland gives off a scent that attracts other angry bees to the spot, thus pressing home the attack. Observe hygiene by keeping your equipment and bee suit clean and observe oral hygiene, since bees are sensitive to foul smelling breath. Bees can also detect carbon dioxide, so the beekeeper should avoid breathing into the hive and blowing onto a frame of bees.

Sound

Though bees don't hear sound in the air, they can detect vibrations in the hive, including the piping sound made by queens who are on the point of emerging from their cells, and the vibrations made by an excited forager returning to indicate the source of nectar or pollen. The successful forager indicates the direction and proximity of a source of food or water or a potential new location for a swarm by performing a "waggle dance". This was discovered and interpreted by Von Frisch, and is the only animal language that man can understand.

The Artificial Hive

Men still collect honey from wild colonies in jungle trees or from cliffs and caves, risking death from being stung and falling. Bees have been kept since ancient times, hived in all manner of receptacles: clay pots or pipes as in ancient Egypt, or hollow logs suspended above the ground, which is still done in Kenya. In Europe, hives were made of straw and called "skeps". The modern hive has removable frames, an invention of an American clergyman, L. Langstroth. There are many types of modern hive that have removable frames, including the one shown, the British National Hive, most common in the UK. Other common hives, including the Langstroth, the Dadant, and the Commercial, have the same basic design.

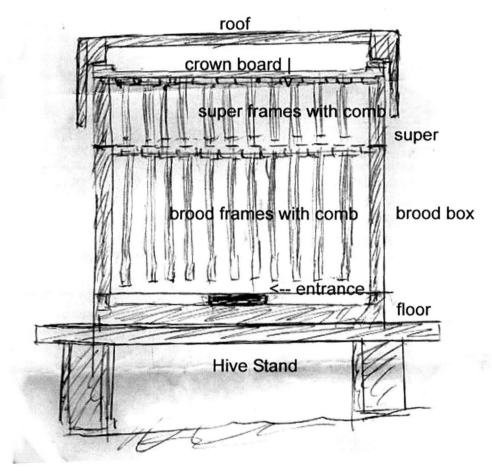

Diagram of the National Hive

The Hive consists of:

a) Brood Box, holding 10 – 11 frames. Outer dimensions 18 inches square, 10 inches deep. The frames hold wax comb, the same combs that the bees make in nature, but held within the frame so that it can be lifted out of the hive for inspection. Though a strong colony can build the comb in an empty frame, the chances are that it will skew away from the plane of the frame and be attached to adjacent frames or the hive wall. So new frames that are introduced by the beekeeper will have foundation fitted – that is, a flat sheet of beeswax that has a honeycomb pattern embossed on both sides. The bees will build their cells on this to make regular comb. The frames usually have metal spacers at each end of the top bar, or else wedge shaped top edges to give 1.4 inch spacing between the centres of each frame, the same spacing that bees observe in nature, to give an ideal space between combs, about 0.4 inch or 1cm - "Bee Space", in which the bees will not build.

b) Super, holding 10 – 11 honey frames. These are the same length as brood frames, but 5.5 inches deep, used for storing honey.

c) Flat Roof

d) Floor, a flat board with a wooden rim about half an inch deep, to leave space below the brood frames. There is a gap at one side of the rim to provide an entrance or fit an entrance block. Some beekeepers use a narrow gauge mesh floor to prevent fallen varroa mites from climbing back into the brood frames.

e) Queen Excluder, a flat metal plate with perforated slots or a frame with a screen of parallel wires that allows worker bees to go through, but bars the queen. This is put between the brood box and the super, to prevent the queen going into the super and laying eggs in the super combs, since they will be used for collecting the honey.

f) Crown Board, usually made of plywood, set within a wooden frame. In the centre is an oblong hole with semi-circular opposite sides, in which a Porter Bee Escape can be fitted for the clearance of bees when the time comes to clear the super of bees in order to remove the honey.

The hive is put on a stand, to raise it at least 6 inches above the ground. A hive stand should be strong enough to bear the weight of a hive full of honey. In a good season the bees may fill up to 4 supers with honey, which if left on top of each other could each weigh over 25 pounds. The total weight of a hive with full supers could be as much as 220 pounds (100kg), so the stand needs to be robust. A pair of breezeblocks on solid ground will make a good stand and allow airflow beneath the floor. The stand should provide a level base for the hive, since a sloping hive will distort the combs.

Crown Board

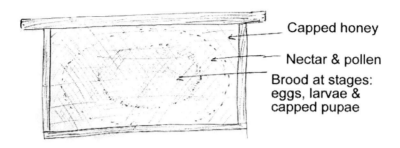

Capped honey

Nectar & pollen

Brood at stages:
eggs, larvae &
capped pupae

Feeder hole and slot for Bee Escape

Top Bar Hives

Today there is a body of beekeepers who are trying to reproduce the natural hive structures that bees make when not constrained by artificial hives. These hives are without frames, but use top bars that the bees are meant to attach comb to. Two popular types are the Warré Hive and the Top Bar (Trough) Hive.

The Warré Hive was developed by Abbé Émile Warré (1867-1951) for the purpose of simple, economical beekeeping. It has cube shaped chambers, on top of each chamber rest horizontal wooden bars. The bees are meant to build comb to hang from each bar. The top chamber is the first to be occupied when a swarm is captured. As the colony thrives and fills the top chamber with comb, brood and honey, a second chamber with top bars is put below it and the colony gradually builds more comb and works its way down into the lower chamber. Then another chamber with top bars is put below that and so on. Eventually the top chamber will be filled with honeycomb and can be cut off and the honey extracted.

The Trough Hive is a long horizontal trough with a triangular cross section fixed to a stand or suspended from the branch of a tree to protect it from ground predators. It has top bars resting transversely across its length and a covering roof that can be lifted off. The intention of the triangular cross section is to allow the bees to build their own comb as they hang in a chain, without the constraint of an oblong frame. The colony expands along the length of the trough by building comb from the central top bars outward to the adjacent bars as they are added by the beekeeper. Honey is taken by removing outer bars of capped honeycomb.

A problem with top bar hives is that the bees do not always follow the lines of the bars: they may build comb at an angle that traverses several bars. This means that the bars cannot be removed separately should the combs need to be inspected for brood diseases. Nor can outside bars be easily removed to get the honey, in the case of the trough hives. To encourage a captured swarm to build comb along the top bars, put a line of wax along the lower edge of each bar, so they can use that as a starter for comb building. It may work, but bees don't always do what they're told.

Starting Beekeeping

The best advice to those who would start keeping bees is to join your local beekeeping association. The British Beekeepers Association can give you contact details for your local branch.

British Beekeepers Association, National Beekeeping Centre, Stoneleigh Park, Kenilworth, Warks, CV8 2LG

It also has a web site. http://www.bbka.org.uk/

In joining a local branch you will be able to observe expert beekeepers in action and benefit from their advice. Other benefits from joining the association are:
- Savings through sharing expensive equipment such as honey spinners and wax extractors.
- A source of bees from swarms and nucleus, far cheaper than buying from commercial suppliers.
- Instruction and assistance in handling bees.
- Insurance for your hives.
- Belonging to a friendly community of people with a common interest.

Collecting the Hive

Often retiring beekeepers advertise their hives in local papers, sometimes with equipment such as smokers, extractors and spare hive parts. The best time to collect a hive is in early spring, when the bees have consumed most of their stores and the hive is light. Do this in the evening or in cold weather, when all the bees are in the hive and not flying. You may need to use a wheelbarrow to carry the hive from its stand in the apiary to the transporting vehicle. To prevent bees escaping, stuff a damp rag or sponge into the hive entrance, and ensure that the floor is firmly attached beneath the brood box. If there is a super on top, make sure that it is well stuck to the brood box. This will usually be the case, since bees stick together any adjoining parts of the hive with propolis, an aromatic resin that the bees make from various plant materials which has antiseptic and anti-fungal properties. The bees apply it to gaps and crevices in the hive to protect the colony from draught and damp. To be doubly sure that the hive will not come apart during the move, bind the hive from top to bottom with a tight strap band.

If the hive is heavy, get someone to help lift it and place it in the wheelbarrow and wheel it from there to the vehicle. Don't forget to bring the hive stand, unless you have already set up a stand in the new apiary. Take the hive to the new apiary and place it securely on its new stand. Finally remove the cloth or sponge from the hive entrance. It is advisable to wear your bee suit, gloves and rubber boots when lifting the hive from the old position and placing it in the new position. Don't forget to open up the hive entrance after you have placed it on its stand.

The best place for a hive is a plot sheltered from wind, without any overhanging branches that may break off and hit the hive. If there are any adjoining fields or paddocks, fence off the hive area with the hive more than 7 feet (2m) away from the fence, so that livestock cannot get too close and knock it over. Place the hive with the entrance facing away from footpaths and ideally facing a hedge or wall that will make the bees fly up and above the heads of livestock or people. It is a good idea to place a sheet of impenetrable material such as corrugated iron or carpet on the ground in front of the hive to suppress undergrowth, which could obstruct the bees' flight to and from the hive entrance.

Urban Beekeeping

There is a growing body of urban and city beekeepers who manage to find secluded plots in gardens, allotments and even sheltered roof tops. There are reports of very productive crops of honey, since the average temperature in cities is higher than in the countryside. There is also a wide variety of foraging in gardens, trees, parks and derelict waste ground, generally free of insecticides and herbicides. See my remarks about the environment in the Annex.

Essential Equipment

1. Hive and hive parts: these are described in the previous chapters.

2. Bee suit: the essential item is a veil, a fine black mesh which covers the face and is fitted at the top to a hood or a hat, and at the bottom to a jacket or a boiler suit. They are usually white, to keep the wearer cool and indicate cleanliness. To complete the outfit, wear Wellington boots and thin marigold or disposable gloves.

3. Smoker: a tin container with a hinged funnel on top and a side tube to admit puffs of air from attached bellows. This is filled with corrugated cardboard rolled into a cylinder, which is lit. By squeezing the bellows, air is puffed into the container and it emits smoke through the funnel, which is directed into the hive to pacify the bees.

4. Hive tool: a piece of flattened metal about 10 inches long and an inch wide, with a sharp straight chisel at one end and a bent scraper at the other. I prefer to use a stainless steel table knife, since it doesn't cut the wood of the hive or the frames and is safer.

5. Honey extracting equipment: spinner, strainer and settling tank. These are described in the later chapter on collecting the honey. They are expensive if purchased new, so it is best to borrow or share with your local beekeepers. Extracting equipment is not needed if you are a "natural" beekeeper using top-bar hives, or are just taking honey off a few frames. In that case, cut the honeycomb off the frame or top bar into a kitchen colander, chop it to break the cells of honey and strain it through a fine mesh.

Managing the Hive

In keeping with my aim to reduce hive manipulations to the minimum essential for productive beekeeping, the following timetable is a guide to managing the hive throughout the season. The main operations are: first inspection, replacing old comb, raising a nucleus, taking the honey, and preparing for winter. The months indicated may vary, depending on the weather and local climate. They are for a temperate maritime climate in the Northern hemisphere, but like the weather, bees are unpredictable, and the timing of these operations will depend on the state of the hive as well as the seasonal weather and location. The months are indicated as a guide, not a hard and fast timetable, so the beekeeper must be observant and alert to the state of the hive and the weather, and act accordingly

January - March:	Check Stores/ Feeding
April on a warm day:	First Inspection
May/June:	Brood Inspection
July/August:	Collecting the Honey
September/October:	Preparing for Winter

Safety Precautions

Use common sense! Don't open the hive when your near neighbours are out in the garden at the weekend.

1. Only open up a hive on a warm day, when bees are flying in and out of the hive in large numbers. There will be less disturbance and cooling of brood.

2. Ensure that your bee suit has no gaps. In particular, that there is no gap between your gloves and the sleeve, and the trousers are firmly tucked into the top of your rubber boots. The veil or mask in front of your face should have no holes and it should not touch your skin.

3. Use well-fitting rubber gloves, either standard household gloves, or if you are confident that the bees are relatively docile, disposable surgical gloves, which will be more hygienic for the bees. I would not recommend the leather gloves with canvas sleeves, sold by beekeeping suppliers, since they make delicate handling of frames and bees difficult. Clumsy handling upsets the bees, and the leather doesn't stop the sting penetrating to your flesh: in fact it often provokes stinging.

4. The smoker should be well lit and charged with cardboard or sacking.

5. Have a clear plan of what you intend to do: the manipulations that will be performed and equipment such as supers and comb or foundation that may be needed.

6. Don't run a petrol mover or strimmer near the hive.

Checking Stores

Assuming your hive was collected in winter or early spring, the first thing to do is to check whether there is enough honey left in the combs above the clustered bees. This can be checked by weighing the hive, but that is a difficult operation, so a fair estimate can be made by tilting it slightly from behind. Be careful not to topple it!

A better way to check is to take off the hive roof and look through the feed hole in the crown board. Do this on a cold day or towards dusk, when the bees are not flying, and be sure to wear your bee hat with the veil fastened. You should be able to see capped honeycomb on one or two of the frames near the feed hole: you can confirm this by probing with a length of stiff wire to check for capped honey. Be careful not to disturb any bees that may be clustering up to the feed hole.

If you are still unsure about the depth of honey, then place a block of the fondant over the central feeder hole in the crown board. If the bees have consumed nearly all the honey at the top of the hive, they will take down sugar from the fondant.

Fondant for feeding bees can be purchased, or you can make it by boiling a quarter pint of water, adding 1kg white granulated sugar. Heat to 114 degree C until all the sugar is dissolved. Then let it cool and stir in a teaspoonful of icing sugar and pour it into flexible plastic containers to set.

Spring Feeding

The good beekeeper will have left enough honey on the hive for the bees to last until the first spring blossom. If honey was taken off and the hive was still light in Autumn, then it should have been fed sugar syrup in Autumn, to make up sufficient stores for over-wintering: i.e. if the bees had enough honey stored for over-wintering, there should be no need for spring feeding. But if the stores are found to be depleted by mid-March, or early April if the spring is wet, spring feeding will be needed to prevent the bees consuming all their stores. The rate of honey consumption will increase rapidly in spring as the queen starts laying and the worker bees feed the brood, and if the weather is bad they will not be able to bring in nectar from early spring blossom, so they must be fed sugar syrup to prevent starvation.

Sugar syrup for spring feeding is prepared by dissolving 1 kilogram of white granulated sugar (cane or beet sugar) in 1 litre of boiling water. The sugar syrup is put in the feeder, allowed to cool and then put on the top of the hive, on the crown board and above the feeder hole. Make space for the feeder by putting an empty super or two, depending on the height of the feeder, on top of the crown board. Then put the roof on, to fully enclose the feeder. Make sure the feeder doesn't drip too much, or the syrup will pour through the hive and onto the ground, making a sticky mess and waste. This could start hive robbing by other bee colonies, especially in warm weather, and it is a significant risk in autumn feeding.

Check the feeder after a week or so. The bees should have taken most of the sugar down into the hive. If the weather is poor, they may need feeding again. Beekeepers who wish to split the hive in summer, to raise a nucleus colony, will continue feeding so that the raising of brood is accelerated.

First Inspection

When opening the hive, always chose a warm day with little wind, and a time of day when the bees are flying strongly, normally late morning and early afternoon. Spring is the time to open up the hive for the first inspection. Bring a super with comb or foundation, a queen excluder, 2 or 3 spare brood frames of comb (if the hive is old, it may need replacement brood comb) and your hive tool or steel table knife. During the following operations, which require the hive to be opened, stand behind the hive, i.e. with the entrance at the other side so that you don't obstruct the bees' flight.

Clear the debris on the hive floor:

Puff smoke onto the hive entrance, lift off the roof and place it upside-down on the ground beside the hive. Puff smoke over the crown board.

Use the knife to prise the floor from the bottom of the brood chamber, lift the hive off the floor and place it on top of the inverted roof , twisting it at an angle so its 4 sides straddle the sides of the roof, leaving space beneath the brood box.

Scrape the debris off the floor and replace the hive on top.

Check the state of the hive and put on the queen excluder:

Prise the crown board off the top and gently shake any bees hanging on it into the open hive. If there is a super on top of the brood box, smoke it from the top to drive most of the bees down into the brood box. The queen may have been laying in the super, so take care to smoke the bees down into the brood box and it is unlikely that the queen will stay in the super. Using the knife or the hive tool, lever the outermost left hand frame from the super. Apply a bit of smoke to clear the bees from the top of the frame and lever it gently but firmly from either end of the top of the frame using the knife handle to break the grip of any sticking propolis – see figure below.

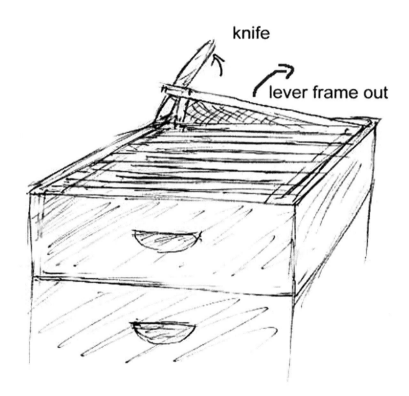

knife

lever frame out

Removing a Frame

Lift the frame on the left hand side of the super, standing from behind the hive with the entrance facing away. Gently shake any bees off it into the top of the super. Check whether this frame has nectar or honey in the comb, and place it inside the hive roof (which is upside-down on the ground). Then lift and check the next frame, which should be easier since you can shift it left into the space left by the first frame. Check this frame for nectar, honey and brood in the form of eggs, larvae or sealed brood. Replace this frame in the super, into the space left by the previous comb, and repeat this for all the frames in the super, working from left to right, then replace the first comb that was taken out into the right side of the super. If there was any brood (eggs or larvae) in the super combs, the queen must have been there, but by now should have gone down into the brood box, driven there by a combination of smoking and shaking bees off the super frames.

Split the super from the brood box by cutting around the hive along the between the super and the brood boxes, and lever off the super. Put the super on top of the inverted hive roof on the ground, and puff more smoke over the brood box if the bees are agitated. Then place the queen excluder on top of the brood box, aligned so that the slots or wires are at right angles to the frames. Place the super back on top of the queen excluder, then the crown board then the roof and the operation is done!

If the hive was over-wintered in a single brood only, then the operation is much simpler. Take off the roof and crown board as before, place the queen excluder over the brood box, then a super with frames of comb on top. If you only have frames of foundation, then that will do, but it means that the bees will take longer to draw out comb before filling with honey.

During the above operations, you should be able to assess the strength of the hive, which will indicate future operations to be planned as the season progresses. For example, a strong hive with an old queen is likely to swarm in the summer. The risk of swarming can be reduced by a number of measures, such as splitting the hive in two, taking off a nucleus, replacing old and damaged comb in the brood chamber with new frames of foundation, providing extra supers so that there is enough space for honey etc. These are covered in the following chapters.

If the hive is strong at this stage, with a lot of honey coming in, instead of putting a super on top of the brood chamber you could consider putting a brood box with frames of foundation on top. This will provide drawn comb later should you decide to split the hive for an "artificial swarm" or to raise a nucleus.

If the hive over-wintered with a super and the frames contain honey from the previous year which has granulated in the comb, or if there is a nearby field of early flowering oil seed rape, which tends to granulate early, remove those combs and replace them with new comb or foundation. The granulated honey can only be recovered by cutting it from the frame and melting it in pan in the oven at a low temperature, about 85 degrees centigrade. After about 90 minutes, the comb will have melted. Let it cool for about an hour and the wax will have set on the top of the pan. Make a hole in the wax at opposite sides of the pan and pour out the honey into jars. Put the wax into a sealed container for rendering later – see wax production.

If the super is nearly full with this year's liquid honey and nectar, don't remove any of that honey, since there may be a cold spell of weather in May and June and the rapidly growing colony will need those stores. Place the new super on top of it.

Brood Inspection

At some stage during the season, you may need to inspect frames in the brood chamber. Since this inevitably causes disturbance, you need to have a good reason for opening up the hive and delving into it. Before you consider doing this, you must check that drones are flying in the vicinity, since there is always a risk that you may harm or kill the queen when you open up the brood chamber. If that happens, then the bees will raise some emergency queen cells, but if this is done too early in the season, before drones have hatched, then the new queen will not be mated and the colony will die.

You can see the state of a hive by observing the entrance and the flight of the bees, then by lifting off the roof and looking at the top of the crown board. This can be done with minimum disturbance and little use of smoke. The conditions that you might find are as follows:

Bees flying strongly, bringing in pollen

– all is well. Lift the roof. If bees are crowded on top of the crown board, and you can glimpse new white wax through the feeder hole, then puff some smoke over it and lift the crown board. Check some frames in the super beneath: if they are mostly full of capped honey, put another super on top. If you are short of spare supers, you could harvest the honey already in the hive, taking the capped frames and replacing them with frames of empty comb or foundation. If there is a good flow of honey, and the hive is strong, it will be worth inserting a few empty frames between the frames of comb and foundation in the super. The bees may build their own comb on these empty frames, which will provide you with cut comb honey, most delicious and a desirable product.

Before you open the hive to inspect the brood, make sure that you have on hand all the items you may need, such as spare frames fitted with foundation. Only do the operation on a warm late morning to early afternoon, when the bees are flying strongly. The smoker should be well lit.

Before opening the hive, puff some smoke in front of the entrance. Don't blow it into the hive, just let it drift in.

Take off the roof and put it upside down on the ground beside the hive. Puff some smoke over the crown board.

Cut around the junction of the bottom super and the queen excluder, then lever the super(s) off and put them on top of the upside down roof, angled so that the super(s) rest on the rim of the roof so that you don't squash any bees.

Take off the queen excluder, shake it gently to let any bees that are crawling on it fall into the brood chamber. Be careful, because the queen may be on it and you don't want her to fall outside the hive.

Puff a bit of smoke over the top of the brood frames to quieten the bees, then lever out the left hand outermost frame, taking care when lifting it out not to crush any bees.

Gently shake the bees off the frame into the brood chamber and examine the comb in the frame, then put it on the crown board over the supers, to remain there temporarily as you go through the rest of the frames in the brood chamber.

Lever off the next frame from the left. This should be first shifted to the left, so that when lifting it there are no bees trapped between it and the adjacent frame. Examine the comb then put it back into the brood chamber, to the left, in the slot that was occupied by the first frame that was removed.

Continue in the same way with the next frame, working from left to right, so that you end up with a vacant slot in the right side of the brood chamber.

When you examine each frame you are looking for the following:
  a) comb with brood at various stages of development: eggs, larvae, and capped brood, surrounded by cells with pollen, then with honey.
  b) any queen cells.
  c) cells showing signs of disease, in particular American Foulbrood.

A frame of capped brood

In the photograph above, you can see cells of capped brood, where the larvae are now in chrysalid form, surrounded by cells of eggs and larva and pollen. On the top corners are cells with honey, some capped with beeswax.

Checking for old or damaged comb

You will need to take off the super and the queen excluder, then take out each frame in turn and examine it. If the brood box appears to be less than half full of bees, with empty frames at the sides, remove an outer frame (in the same way that the first super frame was levered out), and check it and its next frame for damage and uneven sized cells. These will be replaced later with frames of foundation, when the hive becomes stronger and there is a good flow of nectar and warm weather. Put them back and make a note of them (a good idea is to mark the top of those frames with a drawing pin or a dab of pale emulsion paint). Don't delve into the remaining frames in the brood box, since the disturbance and cooling of the brood will set the hive back.

Checking for Queen Cells

If the brood box is full of bees and they are crowded up to the sides, then there is a possibility of imminent swarming. If you want to make sure that they are on the verge of swarming, then check the brood frames for queen cells. These are acorn sized cells that hang down from the comb – see figure.

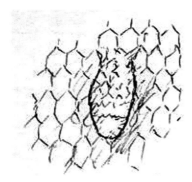

Queen cell

If there are more than 6 queen cells, then the hive is almost certain to swarm, or it has already swarmed. If there is only one or two queen cells and they are near the centre of the brood chamber rather than the outside regions of the comb, then the hive is likely to supersede its queen. That is, a new queen will hatch and the old queen will stay in the hive. The queen mother and her daughter will continue laying eggs until winter when the old queen will usually die. Supersedure is a good thing, for it replaces the old queen with a new queen and you don't lose half your bees and honey production through a swarm.

Before a colony swarms, the old queen will stop laying for a week or so, then leave with the swarm, taking up to half the bees with her. The queen cells will hatch after a fortnight, and the new queens will either fight to the death so that only one remains, or they may fly off, each with a small swarm of about two thousand bees, which is called a cast.

Now the traditional method of preventing a swarm when queen cells have been found is to remove or destroy all the queen cells in the hive. Easier said than done! Almost certainly you will miss some and the hive will swarm anyway. It is far better to split the hive and raise nucleus hives, each with a frame of queen cells.

If the hive has already swarmed (which will be evident by a much reduced complement of bees and lack of eggs in the brood combs), then it is a good idea to remove most of the queen cells in order to prevent possible casts flying off, but leave two queen cells which are still uncapped and have a queen larva with royal jelly.

If no queen cells were seen, but the super is already full off honey, put the additional super with frames of foundation on top of the queen excluder, with the super of honey on top. This will give extra space just above the brood and the young bees will be occupied in drawing out new comb. You may wish to increase your stock later in the season by raising a nucleus, which is described below.

Other operations such as replacing a queen, removing queen cells to prevent swarming, grafting young larvae into queen cells and other specialised queen breeding practices are outside the scope of this book, as indicated in the title.

Replacing Old Comb (May to June)

In the first inspection in April, you will have noted if any of the 2 outer frames have poor quality comb: old black comb, comb that is damaged with wide holes, or comb with uneven sized cells. Drone cells are noticeably wider than worker cells and you don't want the brood chamber to raise a lot of drones. Remove these 2 outer frames and shift the remaining frames to the left side of the brood chamber, leaving a gap for two new frames of foundation to be introduced on the right hand side. Replace the queen excluder and super and crown board and put back the roof. Don't replace frames if the weather is poor and there is little honey flow, since the bees will nibble at the foundation and not draw it out. Work through the frames from left to right. Remove old and damaged frames from the left, shifting the remaining frames from the left to fill the gap. Put new frames with foundation into the vacant spaces on the right. This will establish a regular renewal cycle that will help in preventing the build up of fungal spores in old comb.

If there is some honey in the outer frames that you have removed, then they should be placed flat inside an empty super on top of the crown board, so that the bees can take the honey down. Remove them when they are cleared, after a week or so. The old comb can be cut out of the frame and kept in a sealed container for melting down later and recovering the wax. The old frame can be scraped and cleaned in washing soda, then used for new foundation.

An important point to bear in mind whenever you are lifting a brood fame is to put it back in the same orientation and into the same position relative to its adjacent frames. This is to ensure that the brood pattern and distance between opposite cells is maintained, since it is critical for the bees in controlling the temperature of all the brood.

Checking for Brood Disease

There is one brood disease that is serious and notifiable: American Foulbrood. If you suspect that your hive has it, then contact your area bee inspector to check your hive. If the disease is confirmed, the whole hive and its contents including bees and honey, must be destroyed by closing up the hive entrance, pouring a cupful of petrol into it through the crown board and closing down the roof. The bees will be killed within a few minutes. The hive must then be burned in a pit and the ashes buried.

American foulbrood is suspected if the capping on the brood cells is dark brown and sunken. It is confirmed by poking a matchstick into an infected cell. Instead of a nice healthy grub the cell will contain a sticky brown liquid, which will form a thread when the matchstick is withdrawn. See figure.

American Foulbrood

European Foulbrood is another brood disease, which can kill off a hive. If the outbreak is serious then the bees and comb should be destroyed, but the hive and its parts can be salvaged by applying a blowtorch to the inside boxes and each frame in turn: scald them but don't burn them. Sometimes the bees will overcome the European foulbrood as the season progresses, if the queen is vigorous and the weather is fine, for European foulbrood is a disease of stress.

Other brood diseases that are not so serious are chalk brood, where some of the brood cells have a chalky white filling, and sack brood, where the grub doesn't reach maturity but forms a dry brown sack-like chrysalis lying at the bottom of the cell. These are often a sign of stress in the early part of the season, usually caused by chilling of the brood or nosema (a disease of the gut) and will usually clear themselves in a strong hive with a new queen.

Varroa

Varroa is a small brown mite that grows to about 1.5mm and feeds on the host bee's blood and lays eggs in the brood cells, preferring the larger drone cells near the edges of the frame. The larvae in an infected cell often hatch with deformed wings, and the varroa mite also spreads viruses such as paralysis within the hive and between hives. It was introduced into Southern England in 1992 and has devastated bee hives throughout the country. It was first treated with a proprietary chemical such as Bayvarol, which was successful at first, but because it was not 100% effective the mites that survived became resistant to that treatment. When that and the miticides failed, various organic acids such as formic acid and oxalic acid were tried - hardly a good recipe for "organic" honey! Other thymol type chemicals (Apiguard®, Apistan® etc) in my experiement have aroma so strong that it disrupts communication within the hive, masking the scent of the regulating pheromones. Non-chemical treatment such as forking out drone cells, or shaking icing sugar over the bees so they would clean off the mites, does reduce the varroa count but at the cost of disturbing the colony.

Hives with a lot of crawling bees near the entrance, or bees with deformed wings, are an obvious indication of varroa infestation. With a bad infestation, you can easily see varroa mites on the floor of the hive, which is why many beekeepers now use a mesh screen to prevent fallen varroa from climbing back into the brood chamber. A sheet of white paper can be placed on the hive floor under the mesh for a few days, then removed and examined. The number of varroa mites that have fallen on the paper indicates the degree of infestation. If you examine the fallen mites under a microscope, you may find that some of them have been damaged: bitten bodies or legs removed. These are signs that some bees have developed a resistance to varroa, which is a good development and ought to be encouraged by selective breeding. This is happening in parts of England and it is noticeable that some feral colonies are managing to survive for years and even produce swarms. Such swarms are valuable, and all my colonies have come from them.

Drone with varroa mite on its back

varroa mite

Increasing Stock (May to June)

There are three ways of increasing your stock of bees:
1) collecting a swarm, described in a later chapter,
2) splitting a strong hive in two,
3) raising a nucleus.

Splitting the Hive

If the hive is strong and also if it has more than 1 or 2 queen cells, it may be about to swarm. This can be pre-empted by splitting the hive in two to form what is called an "Artificial Swarm" or "Shook Swarm", a method traditionally used to prevent swarming. There are many articles on different methods of doing this, but I would take the simplest approach, and just put half the brood frames in the new hive, with a dividing frame to confine the bees to half the space, and leave the other half of brood frames in the parent hive. Partition off the empty half of the parent hive with a dividing frame, which is a sheet of plywood or hardboard with same the width as the brood chamber but deeper so that it can partition the chamber into two. If you managed to spot the queen, put her in the new hive. See figure:

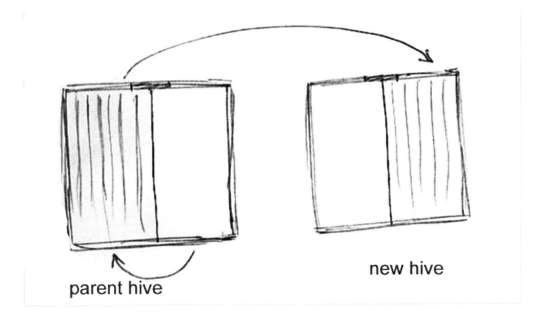

parent hive

new hive

Place the new hive beside the parent hive. If you don't know which hive the queen is in, wait for 4 days or so. If there are few bees flying in the new hive, the chances are that she is still in the parent hive. In that case, swap the positions of the hives so that foraging bees go into the new hive and help the others raise a new queen. After a month or so, you should see bees with pollen flying into both hives, showing that they are queenright. Open up both hives and remove the dividing partitions and fill the empty halves with frames of foundation, or if you have them, frames of drawn comb.

The drawn comb could have been produced before you split the hive, by putting the spare brood box with foundation on top of the brood chamber when you did the first inspection. If you did this, then when you split the hive you would have no need for partition frames. Just make sure that in each half of the split, the adjacent brood frames are kept together: this is to ensure that the pattern of brood is compact so that the cluster can maintain the brood temperature in each hive.

Raising a Nucleus

A nucleus is a small hive, normally consisting of 4 or 5 brood frames in a nucleus box. A standard brood box can serve as a nucleus, with a partition to block off half the space, and a restricted entrance about 0.3 inches by 1.5 inches. The nucleus is used to raise a new colony of bees, or it could be used to hive a small swarm (called a cast). A suitable colony from which to raise a nucleus will have proved itself in the previous year or so. The desirable characteristics are good honey production, non-aggressive bees, and varroa resistance.

At the time for raising a nucleus, the bees should be numerous and at least 8 frames in the brood box should be filled with brood at various stages of development. The operation should be done on a warm day during a spell of good weather, when the bees are flying strongly in late morning. Bring with you the nucleus hive with its entrance closed, and 4 frames of foundation and place it beside the parent hive.

a) Prise the super off the brood box and place it on the inverted roof (as done in the First Inspection, see above).

b) Take the queen excluder off the brood box, shake any bees off it into the top of the brood box. Do this carefully, for the queen may be on the queen excluder. Put the queen excluder on the ground, leaning against the hive.

c) Take out the left hand frame in the brood box. Lift the frame out of the hive carefully to avoid crushing any bees, holding it above the brood box. Shake off any bees into the brood box and examine the frame, check that it has a good depth of honey in the top cells. It should weigh about 3 pounds. If it is empty of honey or eggs and larvae, rest it on top of the super. Lift each frame in turn, working from left to right and examine them, selecting 4 suitable frames. Ideally you want one frame with at least 3 pounds of honey, the others with young brood and eggs, capped brood, pollen and some honey. Put them in the nucleus box, aligned in the same direction that they were in the hive, and the frames with most brood in the centre. That is, 2 frames with eggs and brood will be in the middle: a total of 4 frames.

d) It is important to have some worker bee larvae that are less than 3 days after emerging from the egg. Up to 3 days they are all potential queens, since they have been fed royal jelly. It the bees sense that they are without a queen, they will select some of those young larvae to become queens, and continue to feed them royal jelly and expand those cells into queen cells.

e) Take one or two frames with bees from the hive, and shake them gently over the nucleus so that the bees fall into the top of the nucleus. There should be enough bees in the nucleus to cover all 4 frames. Put the shook frames back in the main hive.

f) Put a crown board and roof on the nucleus. If you are taking the nucleus to a new site, close the entrance, but make sure the nucleus box has a ventilation slot on the top covered with wire mesh, and the entrance is also closed with wire mesh. The slot and entrance should be at least one square inch, otherwise the bees will suffocate. If you wish, keep the nucleus in your apiary until you are sure that it has raised a queen. Open up the mesh entrance after 2 or 3 hours when most of the young bees will have settled into it.

g) Assuming that you didn't put the queen into the nucleus, the colony in the nucleus will soon recognise that it is has no queen and will start to raise 2 or 3 emergency queen cells around newly hatched larvae. If the queen was put in the nucleus by mistake, then the bees will start to raise emergency queen cells in the parent hive. You will be able to tell after a week which colony has the queen, since its bees will be flying strongly and bringing in pollen.

h) After 2 or 3 days when the bees have become accustomed to their new home, feed the nucleus with sugar syrup. Don't do this immediately after making up the nucleus, because it may cause robbing.

i) Do not disturb the nucleus and after 5 weeks it should have successfully raised a queen, who will have mated provided the weather has been good. A good queen will mate with up to a dozen drones in one or two afternoons, until she has enough sperm to fertilise up to a million eggs during the course of her lifetime of up to 5 years. Once mated, she stays in the hive and starts laying within a week or so. She will not mate again and she will only leave the hive when it swarms.

Collecting the Honey (July – August)

In a good year, you may get over 50lb of honey. Normally this is collected after the main nectar bearing plants have finished flowering and are at the seeding and fruiting stages. This will depend on the season and the local flora and crops. In areas where the farmers grow fields of oil seed rape, this could be any time in Spring and Summer, and since the honey from that flower will set in the comb within a fortnight, it is important to take that honey off before it has set otherwise it will be impossible to spin it off the comb.

There are two ways of taking the honey off the hive, either using a clearer board or by lifting each comb in turn from the super. Both methods will need a spare super, either empty or with frames of comb. The clearer board method is recommended if you have 2 or more supers full of honey, but you may need assistance to lift the block of supers off the brood chamber and insert the clearer board underneath them. It has the advantage of clearing the supers of bees, but you have to be careful that the block of supers and the roof are completely bee tight, otherwise robbing will take place since the supers will be undefended when the bees have left. Also, there is a possibility that the bee escape may become blocked and the bees will not clear the supers. If you have only one super of honey, it is quicker to remove the frames from the super and then there will be no danger of robbing.

a) Method 1: Clearer Board

A clearer board is made up of a crown board with slots for one or two Porter Bee Escapes which are fitted into the slots, making sure that they fit well and the rim of the bee escapes are on top. Some crown boards have round holes near the edges, so they will obviously not do as clearer boards.

Smoke the hive entrance, lift off the hive roof and smoke the top of the hive. Place the roof upside down on the ground next to the hive. Insert the bee knife above the queen excluder and cut around the hive then lever the super (or block of supers) off the top of the queen excluder. Lift off the block of supers and place it on the upturned hive roof.

Put the spare empty super on top of the queen excluder, then the Clearer Board on top of the super. Make sure the rim of the Porter Bee Escape is on top, otherwise the bees will not be able to go down through it and into the hive.

Lift the block of supers and place it carefully on top of the Clearer Board, then put the hive roof back on top.

Very important: ensure that there are no gaps between the honey supers, the Clearer Board and the hive roof, otherwise bees from this hive and other hives will enter the supers and take the honey. This will result in robbing and fighting with lots of dead bees. Come back after an hour to check that this is not happening, and if so close any gap with sticky tape. See figure for the arrangement of the hive after putting on the Clearer Board.

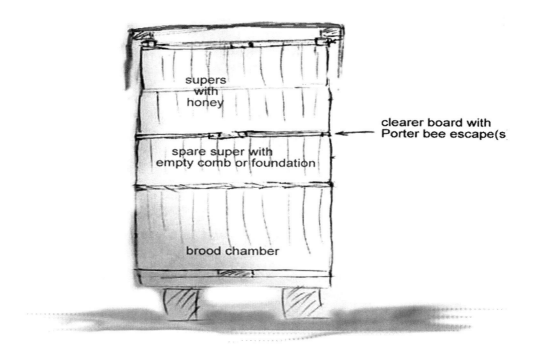

On the next day, bring a wheelbarrow with a sheet of thick polythene spread on top. Lift the hive roof and check that the bees have gone down through the bee escape and the honey supers are free of bees.

Lift off the supers above the Clearer Board and put them in the wheelbarrow. Cover the top with a spare crown board or sheet of polythene. Wheel the barrow out of the apiary and into a bee proof outhouse or utility room, ready for extraction.

b) Method 2: Lifting frames out of the super

Bring a wheelbarrow with an empty super resting on a sheet of thick polythene inside the barrow. Bring also a board or a second sheet of polythene that will cover the top of the super.

Smoke the entrance and lift the roof off the hive and smoke the top and take off the crown board.

Grab a handful of long grass to form a brush. Lever each frame of honeycomb out of the super. At this time of the season, most of the combs will have cells full of honey and capped with wax. If a frame has more than 20% uncapped honey cells, put it back in the super. Otherwise, gently shake off the bees and use the grass brush to brush the remaining bees off the comb, then put the frame in the spare super on the wheelbarrow. Cover the wheelbarrow super with the spare board or polythene sheet.

Repeat this until you have taken all the frames with capped honey off the super. Then put back the crown board and hive roof and wheel away the barrow with its super of honey.

Despite brushing bees off each honeycomb, there will be a few bees inside the spare super on the wheelbarrow. At some distance from the hive (about 50 yards or so), lift the board or sheet off and let those bees fly off, then cover it again. Take the barrow to your bee-proof outhouse or utility room.

Extracting the Honey

Do this in a bee-proof room: an outhouse, utility room or garage. Ideally there should be a high window in the room, to which any bees that have remained in the supers will fly and you can open it and let them out. But make sure you close it again, otherwise you will be invaded by bees seeking to recover their honey! If using the kitchen, it is a good idea to spread a large sheet of polythene on the floor before you start, to collect any spilt wax or drops of

honey. To extract the honey, you will need the following equipment, which most local associations can lend to their members, with advice and assistance:

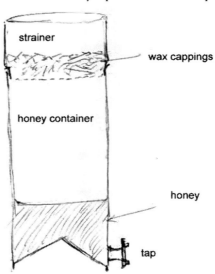

a)   Honey Spinner or Extractor
b)   Honey Container with a tap, on top of which is fitted a Honey Strainer fitted with a mesh that filters out the wax
c)   Uncapping Knife: this is a heavy double-edged knife with which to slice the wax off the top of the honeycomb before putting the frame in the spinner.

I prefer an ordinary sharp carving knife, since I find the specialised uncapping knife unwieldy and dangerous.

This is what you do:

Take each frame in turn, hold it over the Honey Container with its Strainer on top, then slice through the wax capping on each side of the frame, letting the cappings fall into the strainer. Put the frame in the spinner with one face propped up against the mesh. See figure:

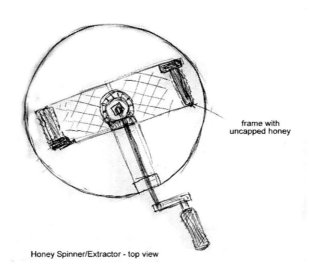

frame with
uncapped honey

Honey Spinner/Extractor - top view

Some spinners will take 2 frames, others 3 and 4. Try to ensure that the frames are roughly the same weight, to balance. At first, turn the spinner gently until most of the honey has spun out from one side of the frames. Then lift the frames and turn them round so the other sides are propped against the mesh, and spin again. When most of the honey has spun out, turn the spinner faster for a minute or so, then lift and turn the frames again and give a fast final spin. The reason for the initial gentle spinning is to avoid the comb breaking under the weight of the centrifugal force on the honey.

When you have spun the honey out of the frames, put them back in the super, which should be placed on top of a sheet of polythene to collect any honey that drips off them. When you have finished spinning, or when the honey in the spinner has reached the level of the frames' top bar (which will retard any further spinning), lift the spinner onto a bench and put the Honey container with its strainer under the tap of the spinner and open the tap of the spinner, so that the honey flows out of the spinner and through the strainer into the container below. Make sure the tap of the container is closed tight! Leave them for at least 12 hours for the honey to drain out of the spinner, through the strainer and wax cappings, and let it settle in the container. Then it can be tapped off into your honey jars.

The wax cappings contain a fair amount of honey even after they have drained. This can be separated by melting the cappings gently in an oven, in the same way that honey that is set in the comb such as ivy or oil seed rape honey – see below. Do not mix this with the honey that you have tapped off already, since its flavour, though quite palatable, will not be as fine as the unheated honey.

Meanwhile, take the supers with the frames that have been extracted back to the hive in your wheelbarrow, still with the polythene sheet underneath and another on top. Smoke and lift off the roof, then remove the Porter Escape(s) from the top board and place the supers on top, making sure there are no gaps. Replace the roof. The bees will move up through the crown board and clean out the remaining honey from the extracted combs. After a week or two the bees will have licked the honey off the wet combs and taken it down to their brood chamber for winter stores. The top supers will be empty of bees after a month or so, so they should be taken away and stored in a mouse proof room for the next season. Don't do what one beekeeper's widow once did: after extracting the honey, she washed the combs under the kitchen tap. So much honey wasted, when it could have been returned to the bees!

There may be a strong flow of honey in August if your hive is near heather, or in September and October from ivy flowers. In both cases, it is unlikely that you can spin out that honey. Heather honey is a gel, and ivy honey sets hard, so the honeycomb has to be cut out of the frame. For heather honey, the combs must be pressed to squeeze out the honey. For ivy honey, put the honeycomb in an oven container, set the oven to a temperature about 85 degrees centigrade and the honey and wax will melt. Let it cool so that the wax sets on top of the honey, then cut two holes in the wax at opposite sides of the container and pour out the honey through the strainer into your honey container.

Both heather honey and ivy honey have distinctive flavours which are highly prized by some, but are not to everyone's taste. Ivy honey will mellow after a year or so and can be blended with other types of liquid honey to make excellent creamed honey.

Some beekeepers put frames without comb or foundation into the supers during a strong honey flow. The bees will make their own comb inside the frame, which instead of spinning out the honey, the comb can be cut out into 6 or 8 ounce blocks and put into plastic cartons with transparent lids and sold as cut comb. This will command a higher price and the flavour of cut comb honey is far more intense than the extracted honey, since the essential oils and distinctive flavours are locked into each individual cell. It is perfect for eating on brown bread toast, since the wax (which is good for the stomach) will be ingested with the toast and not stick in the teeth.

Honey is a clear liquid when first bottled. It consists mainly of sugars which have been processed by the bees from the nectar, to form a combination of fructose and glucose. There are also traces of the aromatic compounds that come from the flowers that the bees have been foraging on, which give various types of honey their distinct flavours. The bees have evaporated most of the water off the nectar to concentrate the sugar and convert it from fructose. They add enzymes that help preserve the honey and inhibit fermentation. Ripe honey should have less than 20% water before the bees cap it, which makes it viscous at room temperature. Indeed, one of the criteria for judging honey is the viscosity and concentration.

The colour of liquid honey can vary from a pale parchment, through amber of different shades, to a dark treacle colour. Most liquid honey will crystallise after a few months and become opaque set honey. See the photograph on the front cover for examples of the various forms that honey takes. Crystallised honey can be melted if you place the jar in a pan of warm water, 90 deg. C, but be careful not to heat it too long for it will lose much of its nature and flavour.

Preparing for Winter (August – October)

The queen excluder must be removed before winter. This can be done after you have taken off the honey, but you may decide to keep it there with a super on top if you expect to get more honey, for example from the heather in late summer. If you are taking the hive to a new location, such as heathland or heather hills, then move them after dark when all the bees are in the hive. The hive must be moved at least 3 miles, otherwise the foraging bees will return to the location you moved it from. If the distance is greater than 3 miles, they will settle after a day or so familiarising themselves with their new location.

You may leave one super on the hive if it has some honey in the frames that is either uncapped or not enough to make extracting it worthwhile, but generally there should be enough room in the brood chamber for all the bees to over-winter. This is the time that most beekeepers feed sugar solution, and various treatments for varroa, nosema and other remedies. As explained earlier, I don't put chemicals in my hives, relying instead on hygienic bees to counter the varroa by grooming. To help them I put a spare empty super beneath the brood box, with a queen excluder between it and the super. This also prevents the invasion of mice into the hive. Many beekeepers use a mesh floor in order to monitor the drop of varroa.

To prevent invasion by mice or wasps, then put a narrow entrance block in the hive entrance, to restrict the gap to about a quarter inch high by 3 inches wide. Wasps are a particular nuisance in late summer and autumn and can kill off a hive unless the entrance is small enough to be defended by the guard bees. A wasp trap near the hive will help, but don't bait it with honey! I use apple juice, jam, marmalade or a combination.

Mice will enter a hive in late autumn or winter when the bees are quiescent. They will make a nest in the outer combs and destroy a lot of the comb, eating some of the stores of honey and pollen. A colony will often die as a result. One way to prevent this is to fix a mouse guard, a metal sheet with small holes, on the hive entrance. The holes are supposed to be large enough to admit the bees, but small enough to stop a mouse. In my experience a determined mouse will get through somehow. I use the queen excluder placed directly below the brood chamber to exclude mice. One old beekeeper used to say that if you could insert an HB pencil into a hive, a mouse will get in. I suppose the same would apply if you checked with a BB or H pencil!

Wax moth is common in empty hives or hives with a lot of old comb and a week colony. There are two species of wax moth: large and small. The small wax moth eats wax on capped honey. The large wax moth eats mainly old brood comb and pupates in silk cocoons around the edges of brood frames and in crevices in the hive. Wax moth can destroy stored brood comb over winter, but this is not necessarily a bad thing, for old brood comb can harbour diseases such as nosema and European foulbrood. There are chemical remedies for deterring wax moth, but as I say in the introduction to this book I believe they harm the bees, for both bees and moths are insects. If you want to store comb honey for more than a few months, it is best to put it in the freezer. This also has the benefit of stopping the honey from crystallizing in the comb.

Ventilation

A lot of colonies succumb to damp in winter, rather than cold. This is because poor ventilation will allow mould and fungal infections develop inside the hive. This is less likely to happen if a mesh floor is used instead of a board floor, but some may be worried about very cold winds chilling the colony. Some crown boards have a series of holes near the edges to allow ventilation, or you could put matchsticks below the corners of the crown board.

Autumn Feeding

A good beekeeper should leave enough honey in the hive for the bees to over winter, without the need to feed sugar syrup. There is no doubt that honey is a superior food, and the growing body of Natural Beekeepers are finding that bees do not suffer from gut ailments such as nosema as much as those fed sugar. However, if by October there is little honey remaining in the hive (which you can tell by weighing, see "Checking Stores", you will need to feed sugar.

To prepare the autumn feed, put a measure of granulated white sugar, either cane or beet sugar, into a container. For every kilogram of sugar boil about one pint of water and pour it over the sugar and stir it until all the sugar is dissolved. When it has cooled, take it to the apiary with a feeder and an empty super or two (depending on the height of the feeder). The sugar syrup for autumn feeding is more concentrated than that used for spring feeding, since the later is consumed immediately by the bees for brood rearing, while the autumn sugar will be stored to over-winter and should be more concentrated to avoid fermentation. There are various proprietary feeders, but I find that a 4 to 8 pint plastic container with about 50 pin holes in the lid is quite satisfactory.

When you feed the bees, you should wear your bee costume and veil and boots, even though you are not opening up the hive, since the bees may fly at you through the feeder hole in the crown board when you lift off the hive roof. There is no need to smoke, unless the hive is very aggressive. In that case, the queen should be replaced: easier said than done! But late summer or autumn is the time to do it. See the next chapter.

Combining Hives

The replacement of a queen is a tricky operation. It is usually done in late summer, when a hive has a queen that is failing due to old age, or the bees are very aggressive. It involves finding the old queen and killing her, then introducing the new queen. It is also risky, in that often the bees will not accept the new queen, so you end up with a queenless hive. In late summer, it is unlikely that the bees will successfully raise their own new queen, so the hive goes into autumn queenless and there are no bees left in it after winter. A far better approach is to merge two hives. If you know that one is too aggressive, then by all means kill its queen if you can find her before you merge. Otherwise, the more vigorous younger queen will survive the merger and you will end up with a strong hive going into winter, from which you could raise a nucleus the following spring.

If the 2 hives to be combined are in the same apiary, the weaker hive should be moved next to the stronger hive before they are combined. This may take a few days, depending on how far they are separated, for if a hive is moved more than 3 feet in one day, the foraging bees will return to its old position and not find the new position of the hive. If the hive is moved less than 3 feet, the foraging bees will eventually learn the new position. On the other hand, if the hive is moved more than 3 miles, then the foragers will immediately adapt to their new location. So you can work out the best approach, depending on how far apart your two hives are in the apiary: if more than say 20 feet, it may be best to move the lighter hive to another apiary and leave it there for a couple of days, then bring it back to your apiary, sitting beside the other hive.

To combine the hives, lift off the roof and crown board from the stronger hive, apply smoke from the top and place a single sheet of newspaper to cover the top. Sprinkle a few drops of mint flavoured sugar solution over the middle of the newspaper, then lift the weaker hive off its floor and place it on top of the other hive, with the newspaper in between. The purpose of the newspaper it to prevent the bees from the hive below suddenly encountering those from the hive on top, before they have had a chance to mingle gradually and become familiar with the different odours. The minty sugar solution will further mask the difference and so reduce the likelihood of the bees fighting. A few puffs of smoke before they are combined will also help to inhibit any aggression. After a day or so, the bees will have bitten through the newspaper and will be fully integrated in the combined hive.

Collecting a Swarm

Swarms generally happen from May to July. A swarm will leave the hive on a warm day, usually from late morning till the afternoon. A prime swarm consists of about twenty to thirty thousand bees, which includes the old queen from their parent hive. The bees leave the hive and fly to a nearby tree and settle in a tight cluster on a branch. The cluster is usually the size and shape of a rugby ball, on average 18 inches from top to bottom. There the bees wait, while scout bees will leave the cluster and seek a suitable cavity in which to establish the new colony. This could take any time from less than an hour, to several days, and this is the time to capture the swarm.

To collect the swarm you need a suitable open container, such as a cardboard box about 14 inches square: a grocery box or 6-bottle wine box will do. Some beekeepers use a straw skep. You will also need an old cotton bed sheet if you need to transport the bees some distance from where they were collected. Put on your bee suit, including Wellington boots and rubber gloves. You don't need the smoker. Usually the swarm will have recently left the hive, so the bees will be gorged on honey and reluctant to sting.

Hold the box just under the swarm and gently scoop the bees into the box. If the swarm is hanging awkwardly so that it is difficult to scoop them with the box, use a flat piece of card to sweep them into the box.

As soon as the swarm is disturbed, many of the bees will fly around. Don't worry, so long as the majority are in the box.

Put the box on the ground, upside down, so that the opening is resting on the ground, and prop up one edge about an inch with a stick or stone to allow the bees to fly in and out of the box.

You will notice that any bees left on the branch will gradually fly off and enter the box. If they don't, they will return to the branch. If the branch is thin enough to be cut off with secateurs or a lopper, then it may be easier to remove it and take it down to the ground and put it and the swarm inside the box before inverting the box. After an hour or so, the bees will be in the box, clustered and hanging on to the inside. It may be best to wait till the evening to give all the bees time to settle.

Spread out the sheet on the ground beside the box, gently lift the box and put it on the sheet, still upside down. Lift the corners of the sheet and knot them together to fully enclose the box and the bees. You can now safely move the bees to the new hive, but make sure the box is steady in the car and upside down with the cluster quietly hanging inside.

Take the bees to the new hive. There are several ways that they can be introduced into the hive:

a) With frames of comb or foundation in the hive, put the box down in front of the entrance, then unknot the sheet and lay it out up to the hive entrance. Lift the box sharply and the clustered bees will fall onto the sheet and start to walk up to the hive entrance and go inside.

b) With frames of comb or foundation in the hive, open the top of the hive and take off the crown board, top super and queen excluder (a super may be needed as well as the brood box if the swarm is very large). Dump the bees from the box into the top of the hive, then gently rest the crown board on top. There will be bees underneath, so the crown board will be resting on the cluster, but as the cluster goes down into the hive the crown board will settle to rest on top of the hive.

c) Put the box with the swarm into an empty hive that is without any frames. Keep the box inverted, with an edge propped up, and close the hive with the crown board and roof on top. Leave it like that for a few days until you can see the bees taking in pollen, indicating that they have a laying queen and are building comb. They will be unlikely to abscond then. This is the best way of ensuring that the bees don't leave their new home, but it has the disadvantage that they will have built some comb in the box, which will you will have to cut out and put into frames.

To reduce the chances of a captured swarm leaving its new hive, the bees must feel secure and comfortable. There must be enough room in the hive to prevent overcrowding, and the entrance must be restricted so that they feel that they can defend it. It also helps if there is some old comb there. I have collected many a swarm merely by leaving an empty hive with some old comb in it, and there is a good chance that a swarm will find it and enter by itself – vacant possession! These empty hives are sometimes called "Bait Hives". It is important also not to feed a swarm until you know that it has settled into the new hive, since after having fed, the swarm may have enough confidence to set off again to find a better home.

Small swarm hanging from a tree

Transferring Comb to a Frame

If you have a colony that has built "wild comb", which is comb that is not in a frame and so cannot be manipulated when you inspect the bees, then you may need to transfer it. This could happen if a swarm has entered a hive without a full complement of framed combs or foundation, or you have left the swarm in its cardboard box as in case c) above. To put the combs in a frame, prepare an empty frame by tying stout cotton threads to the top bar and bottom bar as shown in figure.

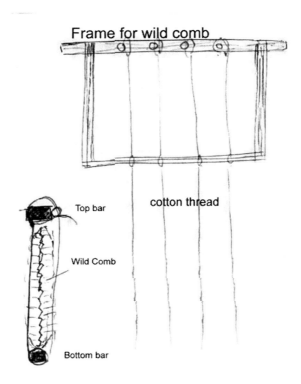

Put the hive into which you are going to transfer the comb next to the box or hive with the wild comb. The destination hive should contain enough empty frames with cotton thread to receive the wild comb, plus frames of comb or foundation to make a full set for the brood chamber: 10 or 11 frames in total.

Lay a frame with the cotton thread horizontally on the top bars of brood chamber, letting the hanging ends of the threads drop over the outside of the hive.

Cut a piece of wild comb out from where it is fixed in the cardboard swarm box or source hive. Use a very sharp knife, to avoid disturbing the bees. Try not to cut into brood cells, and use your smoker to clear the bees from your cutting edge. Gently place the comb inside the frame, to rest horizontally on the tight threads between the top bar and bottom bar. The top of the wild comb should be close to the top bar.

Lift the hanging thread ends over the comb and tie them tight to the top bar of the frame, to firmly hold the wild comb inside the frame. You should be able to lift the frame horizontally without the comb falling out. Put the frame into the brood chamber, and do the same with all the wild comb. You only need to bother with wild comb that has brood or substantial amounts of honeycomb.

Put the crown board then the roof over the brood chamber. If there are some small wild combs with honey still remaining in the swarm box, put an empty super on top of the hive and put those combs inside. The bees will take the honey down into the hive and the empty comb can be taken out later and stored for wax extraction.

Brood and honey comb tied into frame

After a day or so the bees will have fixed the comb in the frame and bitten away the cotton thread.

Colonies in Buildings and Trees

Sometimes the beekeeper will be asked to remove a colony from a building. Usually it will be in a chimney, roof or cavity wall.

It is almost impossible to transfer the colony to a hive by cutting away the comb from the fabric of the building and taking it away in a box, for most of the bees will fly from the comb and return to same place in the building, even if all the comb is removed. You will end up with a sticky mess of brood and honey comb in the box, with dead or dying bees. Those that escape will re-colonise the building and either restore their nest or die off over winter.

There are ways of smoking out the bees, or drumming them up a chimney into a skep. Some books tell you to put a one-way bee escape such as a cloth funnel over the opening where the bees enter and leave the building, with an empty hive on a platform next to it, in the hope that the bees will occupy the hive. However you will need to provide a new queen for the empty hive, for the queen in the building will not venture through the trap into the new hive. Also, it is likely that the bees will find an alternative entrance to their old cavity. I advise you to decline to remove an established colony from a building, and leave it as a good source of possibly varroa resistant swarms.

If the colony is in a tree, the same argument will apply: leave well alone. If the tree has been felled, then it may be possible to hive it by sawing the trunk above and below the colony, fixing it upright on the ground. Ensure that there is a clear passage for the bees to fly in and out of the top of the sawn off tree trunk, either by sawing off the top bit of the trunk just above the hollow where the bees live, or by chiselling or drilling a hole through the top. Then put a crown board on top, and then put a brood box with frames of comb or foundation on top of the crown board. Make sure the whole arrangement is secure and stable. Block off any side to the trunk and with any luck over a few seasons the colony will migrate up to the empty hive. Failing that, they may swarm or in a good season even put honey in the hive!

Wax Extraction

The most efficient wax extractor is a solar melter, which is a box with a double glazed top lid and inside it an open melting pan, such as an old baking dish, with a mesh strainer fitted across the middle. On a sunny day, place the solar extractor in the sun and prop it up so it faces the sun and put the cappings and old comb into the pan above the strainer. The wax will melt and run through the strainer and collect at the lower side of the pan. It can be poured into a mould or collected in the evening when it has set. Make sure that the glazed lid of the melter fits well, since bees will be attracted to the smell of the honey in the melting wax.

For large quantities of wax, a steam extractor may be used. This has an electric immersion boiler on top of which sits a stainless steel container with another perforated stainless steel container fitted inside it. The wax cappings are put into the perforated container. The steam melts the wax and leaves a residue of old brood cells and pollen pellets in the perforated container. The molten wax flows out through a tube, to be collected in a suitable vessel. Your best plan is to borrow a steam wax extractor from your local association.

Uses of Beeswax

Beeswax is a valuable material. It was used for church candles, burning with a bright steady flame and not guttering and smoking like tallow. It is also used for furniture polish, giving a pleasant aroma to the wood, and for strengthening button thread. It is also an efficient waterproof for cloth.

The beekeeper uses beeswax which has been pressed into sheets embossed with a honeycomb pattern for use a foundation, which the bees draw out into comb for brood or honey. I make my own foundation by applying molten beeswax to a damp honeycomb matrix made of plaster of Paris, using a soft wide paintbrush. Though only embossed on one side of the sheet, a strong colony of bees will draw out both sides during a honey flow. Using my own beeswax ensures that it is free of insecticides and other chemicals.

INDEX

ANNEX

Bee Friendly Environment

Post-War agricultural policies have devastated our countryside wildlife. Hedges have been grubbed out in order to extend arable acreage, orchards cut down and the use of insecticides, herbicides and fertilisers renders both arable and pastoral land into a green desert, hostile to insects and the birds and small mammals that feed on them. Indeed, the environment of cities and large towns is more beneficial to bees, and beekeepers in London and Birmingham have harvested more honey than those in the countryside. Too many home gardeners and town councils are obsessed with neatly manicured lawns, closely mown fields and weed free verges. Pollinating insects don't have a chance, and many butterflies and bumblebees are becoming extinct.

Fortunately there is a movement driven by the celebrated gardener Sarah Raven, which is restoring our native meadows and hedgerows. She has persuaded some enlightened farmers to allow boundary strips of uncultivated ground to become meadows. A beneficial side effect is that insect and bird predators of crop pests are allowed to thrive, so reducing the farmers' reliance on insecticides. Some gardeners are now allowing their gardens to grow wild flowers, and so enjoy their natural beauty and butterflies that visit them. They also save petrol and the accompanying fumes, noise and wasted effort of mowing and strimming.

So what are the best plants for honeybees? Starting from early spring, the following are the honeybees' favourites:

Snowdrop, witch hazel, willow, crocus, cherry (not ornamental), all soft fruit (raspberry, gooseberry, blackcurrant, strawberry), dandelion, plum, pear, apple, cotoneaster (an insignificant flower, but the bees go mad on it), clover and bramble, which is the main source of honey in the countryside. All sorts of flowering shrubs, generally with small flowers, that give off significant amounts of nectar. Then the large trees: horse chestnut, lime, sweet chestnut, honeydew secreted by aphids: a lovely dark viscous honey is produced from it. Finally, the autumn flowers – asters, heather, and almost into winter, ivy.

Spare the mower! - wild flowers on uncut lawn

The Joys of Beekeeping

In this short book I have given a detailed account of the essential operations that a beekeeper performs throughout the year. I have outlined the nature and lifecycle of the honeybee, sufficient to give a basic understanding of its behaviour in order to efficiently manage your hives. If you want to read further about the natural history and detailed biology of the honeybee, or on specialised practices such a queen rearing, there are many fine publications available.

There are few moments in life that match the joy of standing near a healthy hive on a warm day, when the bees are flying and bringing in their loads of sweet nectar and colourful pollen. To hear the humming of the hive and smell the intoxicating aroma of the evaporating nectar on a warm summer evening gives a feeling of peace and tranquillity that is rarely experienced in our mundane and hectic modern world: a true communion with nature.

Lightning Source UK Ltd.
Milton Keynes UK
UKOW020636090512

192228UK00001B/1/P